EDUARDO HAD A GRAPE MATH DAY

written by
Lori M. Snell

illustrated by
Courtney Monday

Manufactured in the United States of America

ISBN: 978-1-7323692-1-4

**To My Three Sons,
Joel, Gabriel and Daniel Snell.**

Eduardo hated math! He just didn't understand how to add, no matter how he tried to pay attention in class.

Mrs. Snell was an awesome math teacher. She used all of the strategies that she could: math songs, colored markers, blocks and all sorts of math manipulatives but it was NOT enough to help Eduardo understand how to add.

9 - 7
?
- 1
=
2
+
5 + 5

Today was no different in Mrs. Snell's first grade class. There was Eduardo on the second row from the front of the class, quietly sitting, slumped down in his chair, while Mrs. Snell was showing the class a video about adding.

She always tried to find a short, animated video that would help her students be more engaged in the lesson. The videos would usually show step by step ways to solve a problem, using animal characters, who spoke in the voice of kids. But to Eduardo, the screen was a blur and the voices of the characters sounded like, "Blaaa, blaaa, blaaaa."

School was over, and Eduardo slowly climbed onto his bright, yellow bus where everybody seemed excited and full of life, except for him. Eduardo just stared out of the window as the bus went "bumpity thump" down the road toward his house. All he could think about was the fact that he didn't know how to add. "I'm never going to get out of first grade, if I don't learn how to add," he whispered to himself.

A few minutes later, the bus was stopped in front of his house. Eduardo, grabbed his favorite Super Man backpack and dragged it off of the bus. He was feeling so down that he didn't even hear Julissa say goodbye to him. Eduardo thought that Julissa was the prettiest girl in his class.

He opened the door and dropped his backpack on the floor and went into the kitchen, where he saw his mom putting away fruit that she had just purchased from the fruit market.

She said to him, after she gave him a hug and a kiss on the forehead, "Hello my son. Why do you look like you just lost your best friend? How was your day?"

The big, salty tears began to roll down his soft, brown cheeks. "I don't know how to add and I'm never going to get out of first grade if I don't learn how to add!"

His mother, looked at him, running her hand over his spiky, black Mohawk and said, “Yes you will.” “Come and sit next to me at the kitchen table,” she said.

He sat on the wooden chair, next to her, with his elbows on the table and his hands cupped around his face. His mom said to him, “Let’s eat some grapes.” Eduardo’s eyes stretched wide. He loved grapes. He said with a half smile, “Okay, mom.”

So, his mom took a big bag of smooth, green grapes out of the refrigerator. Then she put a white sheet of paper towel in front of Eduardo and one in front of her. Eduardo's mouth was beginning to salivate. He really wanted those sweet, juicy grapes. But much to his surprise, his mom didn't give him the normal cluster of grapes. Nope, she gave him 7 grapes.

Then she asked him to tell her how many more grapes she would need to add, to get to 10 grapes. Eduardo whined, "Mom! You know I don't know how to add!" She replied softly, "Yes you do. You have 7 grapes, right? Now take more grapes and stop when you say 10, but put the extra grapes on my paper towel."

Eduardo's little hands stretched out toward the grapes and he began to pull and count. He started talking the math through, just like Mrs. Snell had always taught her students to do. She called it "math talk".

He said, "I have seven grapes. I need ten grapes. So, if I add 1 more grape, I'll have 8 grapes. 7+1=8. Eight is not ten, so I'll have to pull one more grape. I have 8 grapes and if I pull one more, I'll have 9 grapes because 8+1=9. Okay, now I'm getting closer. I have 9 grapes, and if I pull just one more, I'll have ten grapes, because 9+1=10."

Eduardo remembered that Mrs. Snell always taught them to recount, to make sure that their sum was correct. So he whispered to himself, “Now let me see. My paper towel has 7 grapes and mom’s paper towel has 3 grapes. When I count-on, starting with seven, I will say 7,8,9,10. Hmm, so 7+3=10.”

Eduardo’s mom reached over with a big, bright smile and gave him a huge hug and said to him, “That’s correct. You did it!” They both laughed. Eduardo was so thrilled that he forgot about how badly he wanted to eat the grapes. All he wanted to do is to solve more grape addition problems. After solving many addition problems, he and his mom sat at the table and ate a bowl full of delicious grapes.

Eduardo was so happy that he now understood how to add numbers together to make other numbers. So, he asked his mom if they could go to the fruit market to buy enough grapes for his class. He said that he wanted to show Mrs. Snell and his classmates to add, using grapes.

His mom thought that was a terrific idea. She said "Let me call Mrs. Snell to see if that will be okay."

Mrs. Snell told her that it would be an excellent idea. So, Eduardo and his mom went to the store and bought enough grapes for his entire class. They separated them into small bags for each student.

The next day, his mom drove him to school to make sure that the grapes arrived safely. Eduardo and his mom took the grapes to Mrs. Snell's class. Eduardo couldn't wait for math time to arrive.

It was finally time for math. Mrs. Snell announced to the class, “My Angels, Eduardo is going to teach us an adding lesson, today.”

Some of the children giggled, because they all knew that Eduardo struggled with adding.

Then Eduardo stood in front of the class and said, “We are going to have a GRAPE day adding! I would like for you all to come up and get two paper towels and one bag of grapes.”

“Grapes!” The children exclaimed.

“Yes, grapes,” Eduardo replied. “We are going to practice solving addition problems with grapes. After we are finished, we will eat our grapes.” The students cheered and clapped.

Eduardo told the students that today they were going to make the number 10 by adding 2 different numbers together. They started with the number 1, and the children used their grapes to count on from 1 to 10 and quickly shouted out the answer, 9. Mrs. Snell asked the children to write the matching number sentence in their math journals.

The children wrote, 1+9=10.
At the end of the lesson, the children had created ten number sentences or equations that equaled 10.

1 + 9 = 10
2 + 8 = 10
3 + 7 = 10
4 + 6 = 10
5 + 5 = 10
6 + 4 = 10
7 + 3 = 10
8 + 2 = 10
9 + 1 = 10

The students had so much fun adding. When Eduardo got home, he ran inside and hugged his mom and said, “Thanks mom! We had a GRAPE math day!”

GRAPE MATH ADDITION LESSON PLAN

Learning Goal

The learners will be able to create various addition math sentences/equations, up to the number 10, using grapes.

Lesson Preparation Time: Approximately 30 minutes
Lesson Duration: Approximately 30 minutes

Materials Needed

1. Grapes (enough for each student to build the target number)
2. Snack size bags
3. Paper towels (2 per student)
4. A math journal or sheet of writing paper
5. A pencil or marker

Directions

1. First tell the students what number (targeted number) will be made (i.e-7, 8, 9, etc.) and write it on the board. The students will write it on the top of their paper

2. Then, the start number will be called (for example: 2)

3. The students will put that number of grapes on one sheet of paper towel, and record that number on their paper

4. The teacher will then ask the students to count on from the first number to the targeted number, using the grapes, but adding them to the second sheet of paper towel

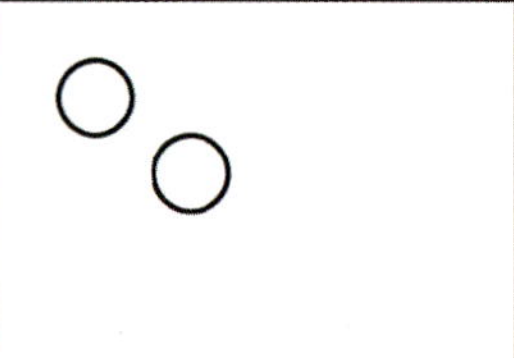

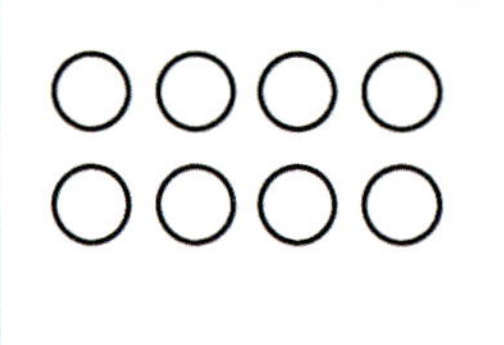

=2+8=10

Lesson Variations: This also can be used to tell addition math stories: Kim had 2 grapes. Her mom gave her 8 more grapes. How many grapes did Kim have in all? Also, this can be expanded into making a number using more than 2 numbers. Add one more paper towel to create 3 digit number sentences,